NERVOUS SYSTEM

Becky Noelle and Priyanka Das

www.av2books.com

Step 1
Go to **www.av2books.com**

Step 2
Enter this unique code
DMULZI135

Step 3
Explore your interactive eBook!

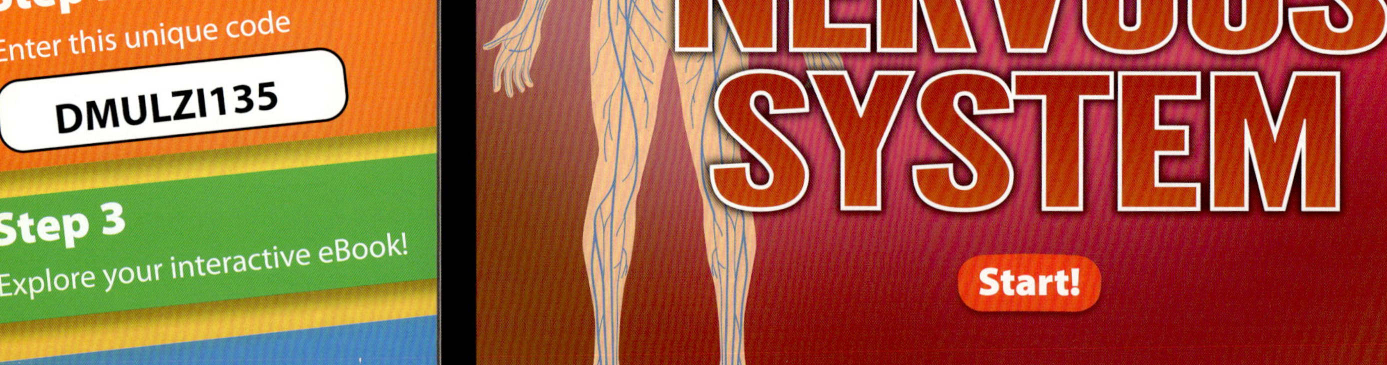

AV2 is optimized for use on any device

Your interactive eBook comes with...

Audio
Listen to the entire book read aloud

Videos
Watch informative video clips

Weblinks
Gain additional information for research

Try This!
Complete activities and hands-on experiments

Key Words
Study vocabulary, and complete a matching word activity

Quizzes
Test your knowledge

Slideshows
View images and captions

View new titles and product videos at www.av2books.com

NERVOUS SYSTEM

CONTENTS

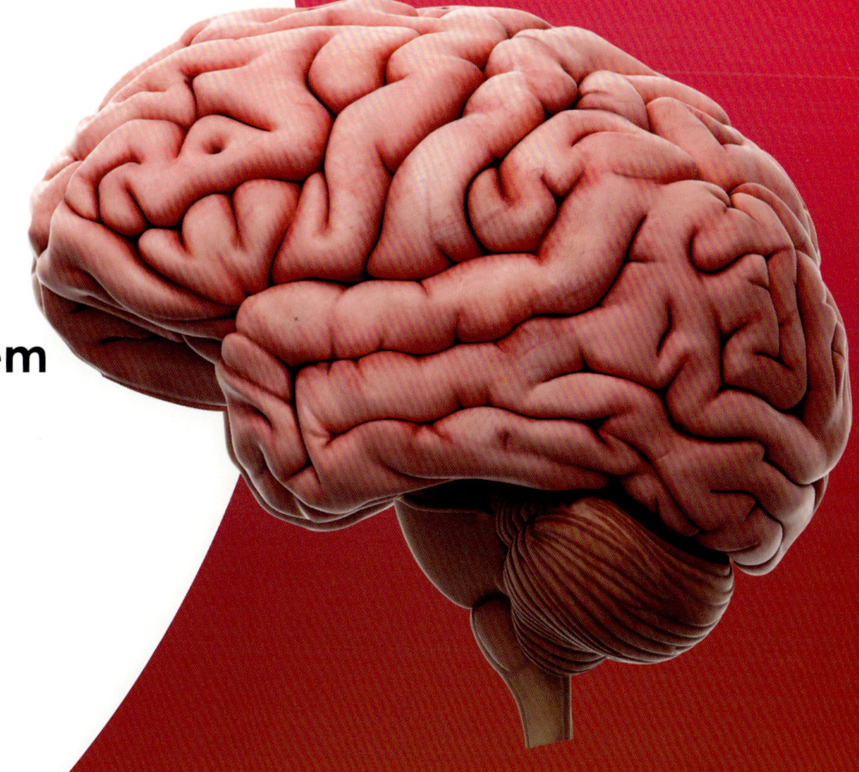

Nervous System

Your body systems work together to let you learn, play, and live. **The nervous system controls all the other body systems.** It senses what is going on around and inside you.

The nervous system carries signals from all over your body to the brain. **The brain responds to these signals.** It tells your body what to do.

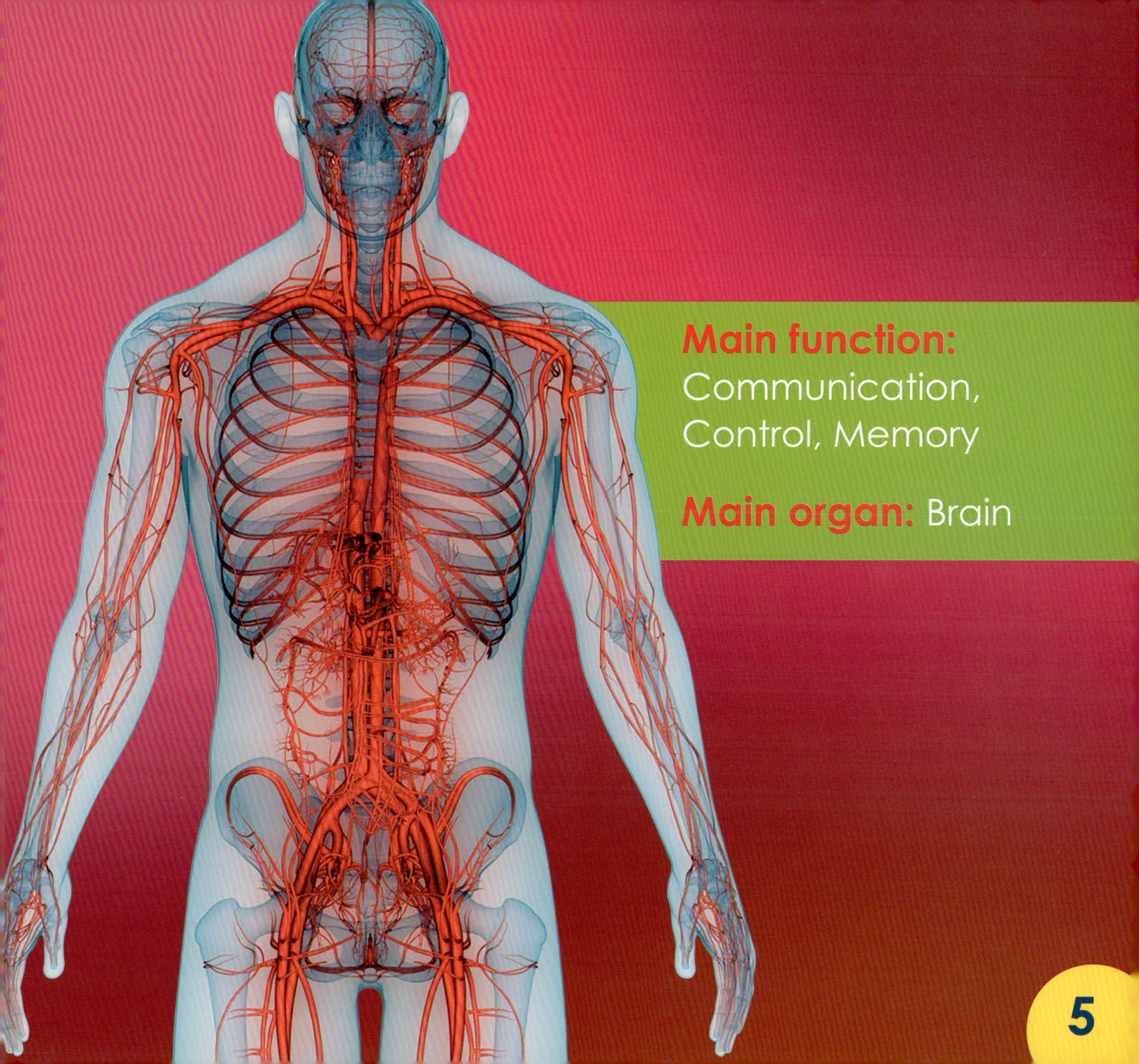

Main function: Communication, Control, Memory

Main organ: Brain

All about the Brain

Your brain is more powerful than any computer. It is constantly getting messages from your body. **You use your brain to think, feel, dream, and solve problems.**

The brain is the most important organ in the body. **It is responsible for storing memories and learning.** The brain also controls movement.

Parts of the Nervous System

Parts of the body called sensory organs take in information from outside the body. **Nerves carry this information to the brain.** Then, the brain tells the body how to respond.

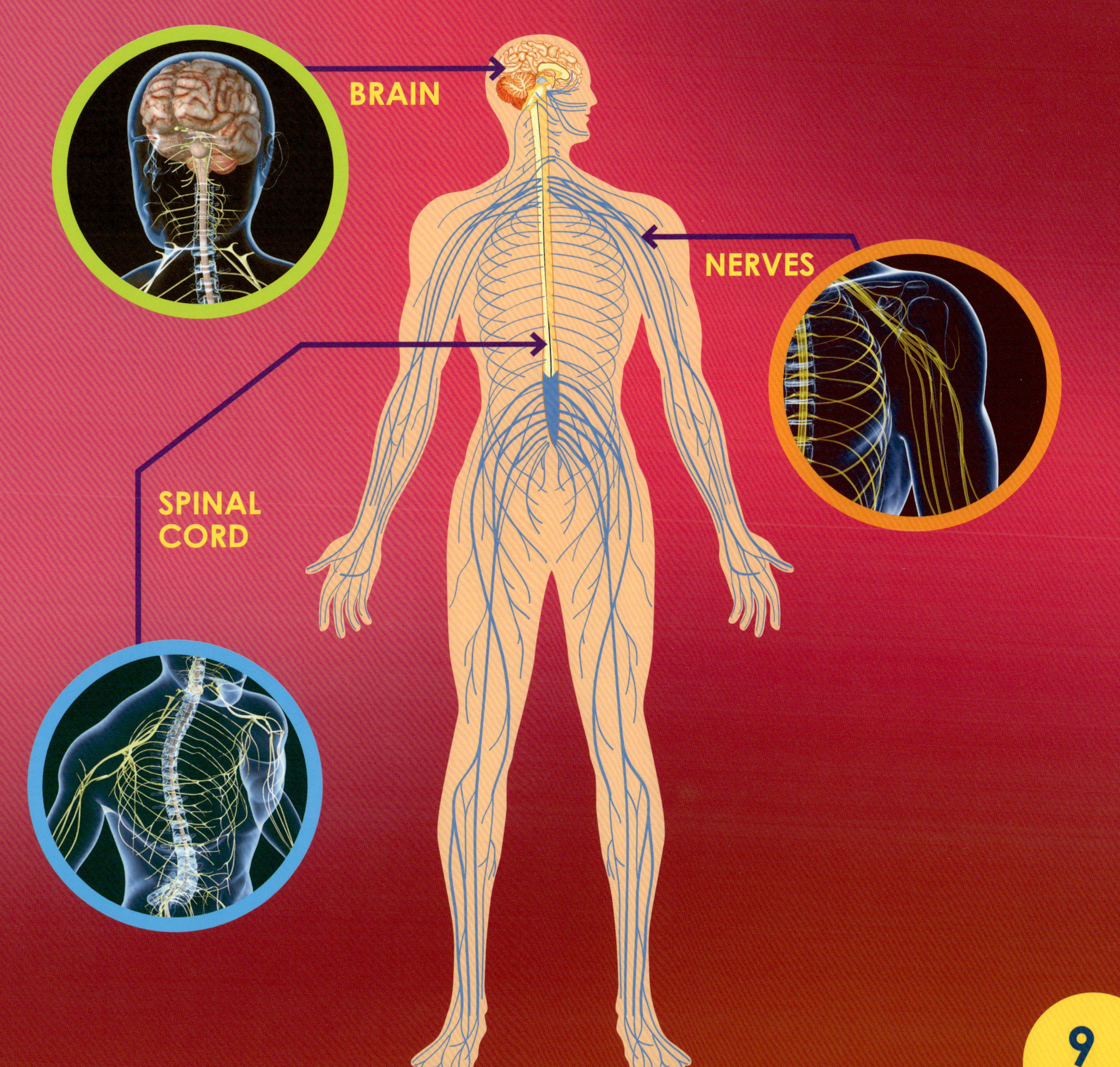
BRAIN
NERVES
SPINAL
CORD

Brain

On the outside, the brain looks wrinkly and pink. The wrinkles make space to help move information around.

The brain is divided into two halves. **These halves communicate with each other.** The right half of the brain controls the left side of the body. The left half of the brain controls the right side of the body.

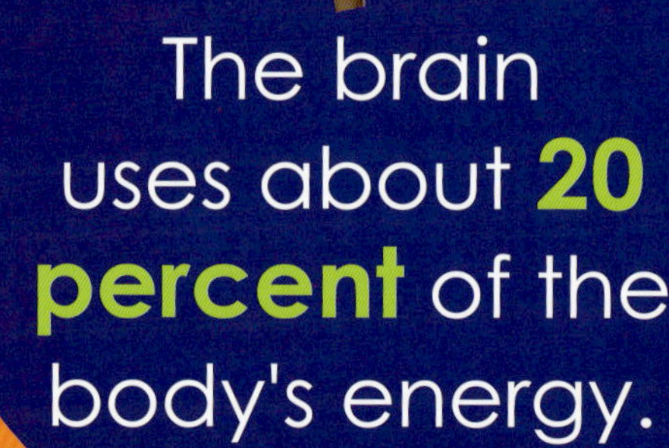

The brain uses about **20 percent** of the body's energy.

Nerves

Nerves carry messages from the body to the brain and back again. **They are made of special cells called neurons.** The human body has billions of neurons.

Nerves can be found throughout the body. Some neurons carry messages to the brain from the sensory organs. Other neurons carry messages from the brain to parts of the body.

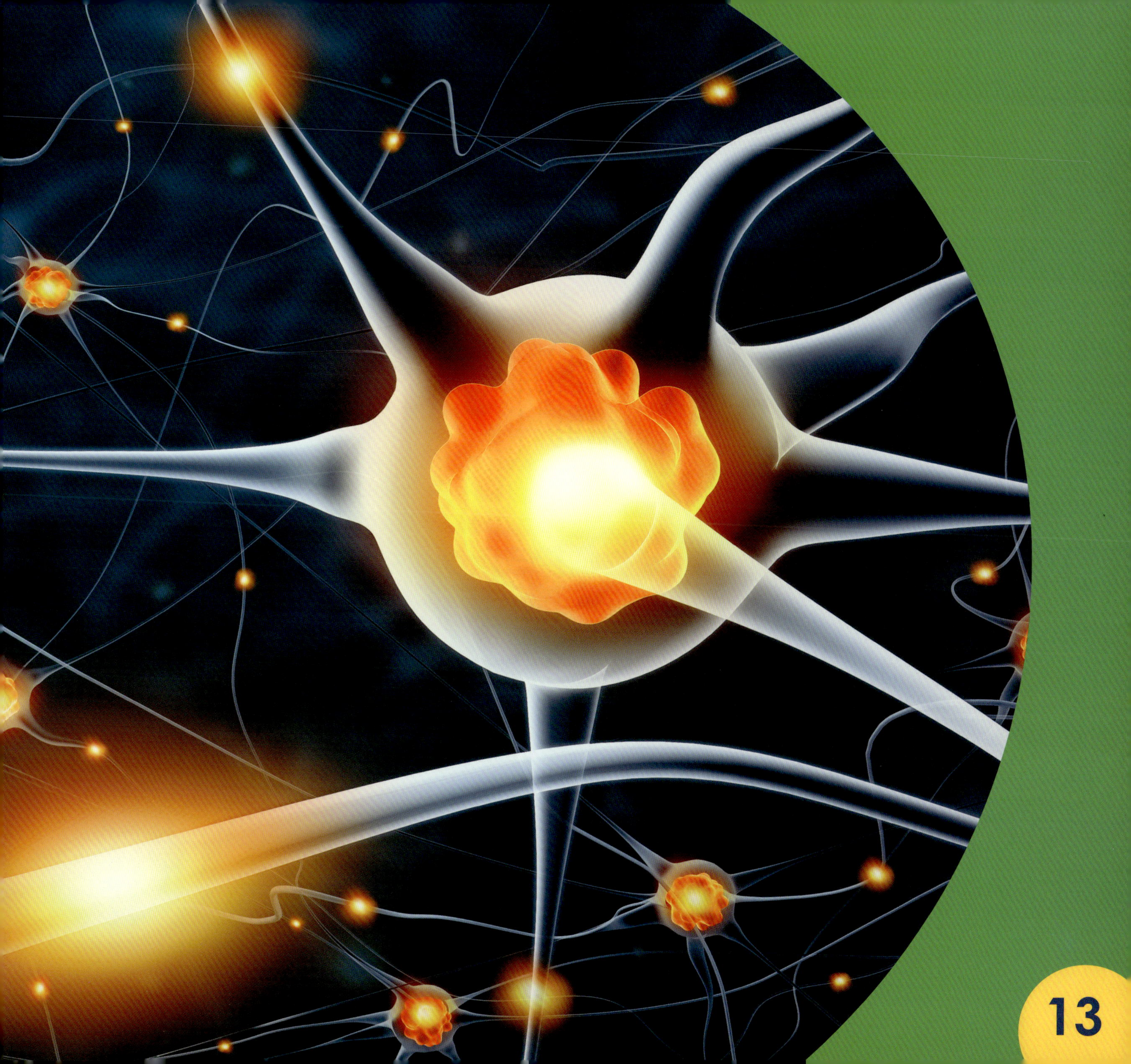

Spinal Cord

The spinal cord is a bundle of nerves. **It connects the brain to the rest of the body.** The spinal cord starts at the back of the neck. It runs down to the lower back.

The spine protects the spinal cord. If the spinal cord is injured, it may lead to paralysis. Someone who is paralyzed cannot move one or more parts of his or her body.

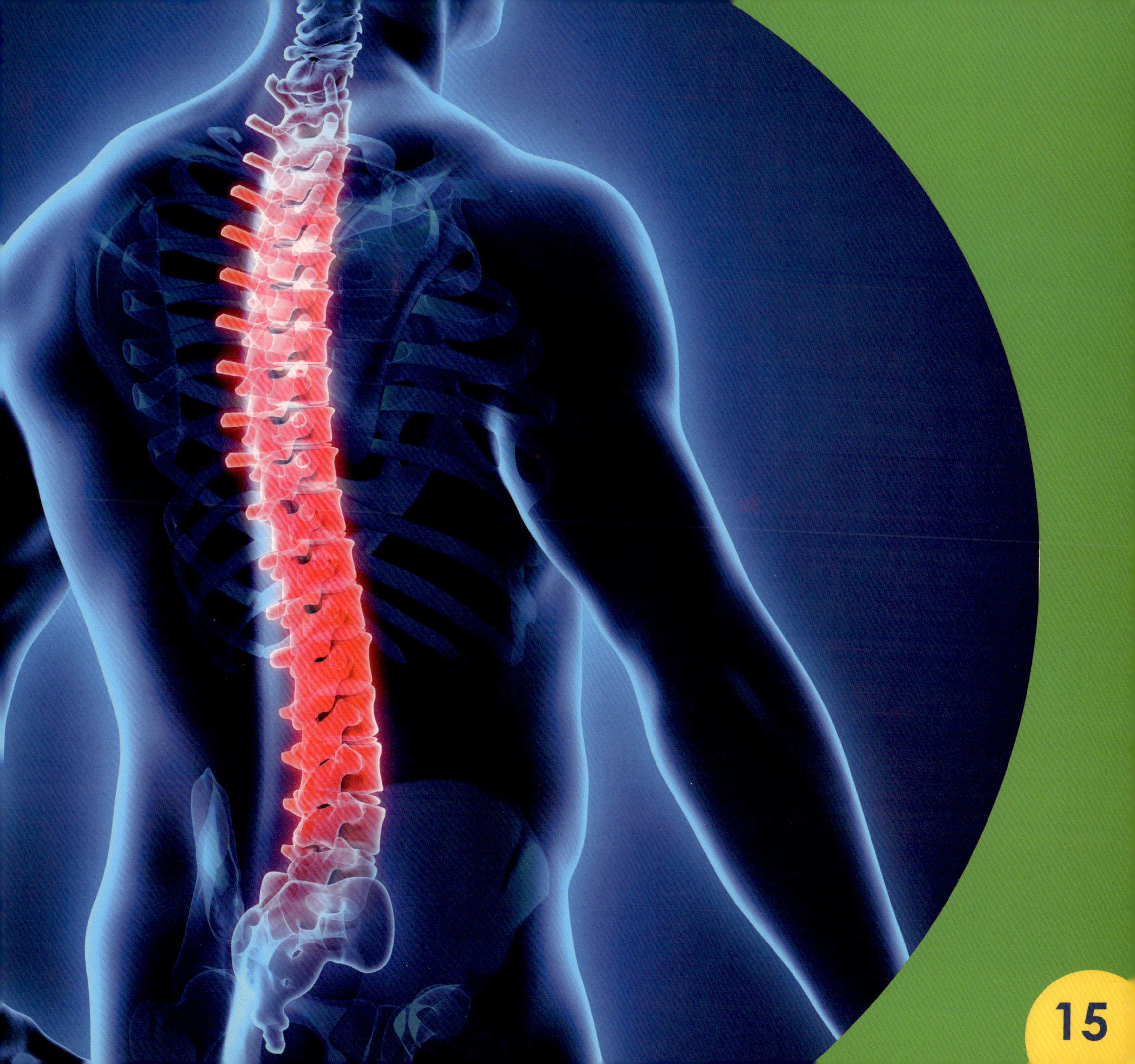

Sensory Organs

Humans have five basic senses. We understand the world around us through them. **The eyes, ears, nose, tongue, and skin are sensory organs.** They allow us to see, hear, smell, taste, and feel.

If you touch a hot stove, the nerves in your skin send a message to your brain. The brain responds, and you feel pain. You pull your hand away from the stove.

Scientists believe **touch** is the **first sense** that humans develop.

Staying Healthy

Water and vitamins help the nervous system stay healthy. Eggs, milk, fish, and meat provide B12, an important vitamin that the nervous system needs.

water

meat

eggs

milk

fish

You can keep your brain healthy by challenging it. **Your brain builds new connections when you learn new things.**

Career Spotlight

A neurologist is a type of doctor. He or she helps people whose nervous systems are not working properly. People may need to see a neurologist if they are having memory problems, bad headaches, or muscle weakness.

Boston Children's Hospital is the best hospital in the United States for children who need to see a neurologist.

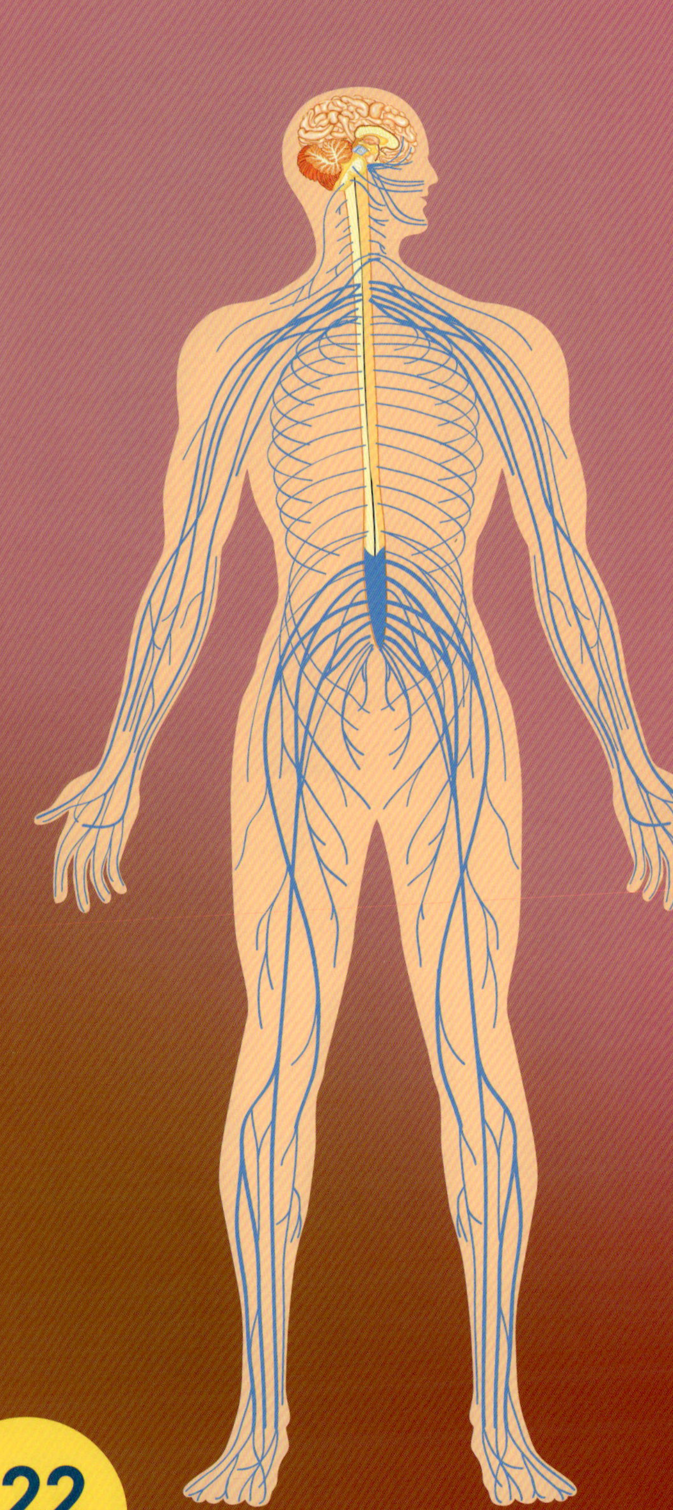

Nervous System Quiz

Can you name the parts of the nervous system shown in these pictures?

A] Brain

B] Nerves

C] Spinal cord

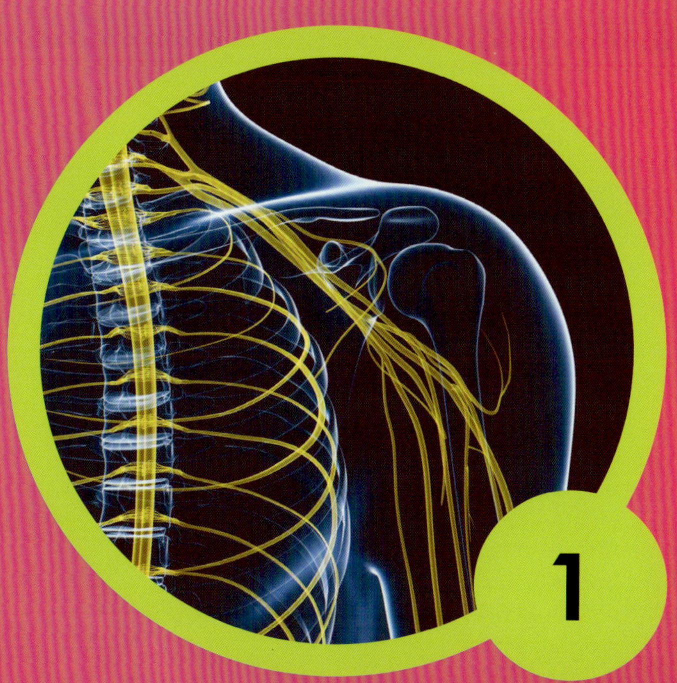

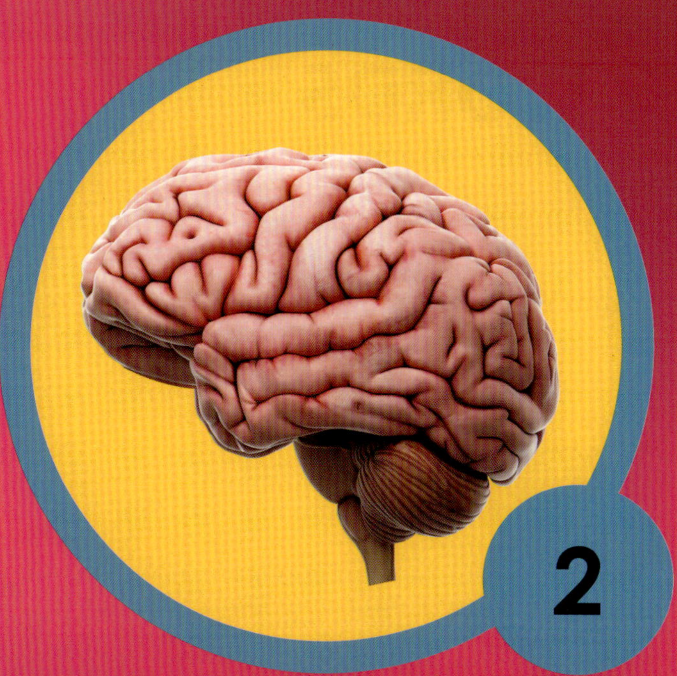

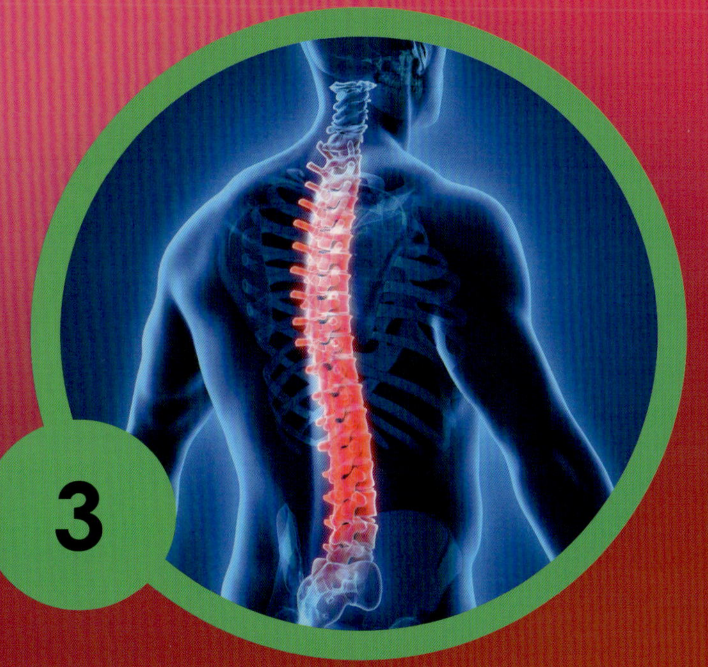

Answers
1] B
2] A
3] C

KEY WORDS

Research has shown that as much as 65 percent of all written material published in English is made up of 300 words. These 300 words cannot be taught using pictures or learned by sounding them out. They must be recognized by sight. This book contains 101 common sight words to help young readers improve their reading fluency and comprehension. This book also teaches young readers several important content words, such as proper nouns. These words are paired with pictures to aid in learning and improve understanding.

Page	Sight Words First Appearance
4	all, and, around, do, from, is, it, learn, let, live, on, other, over, play, tells, the, these, to, together, what, work, you, your
6	about, also, any, for, important, in, more, most, than, think, use
8	carry, how, of, parts, take, then
10	each, help, into, left, looks, make, move, right, side, two, with
12	again, are, back, be, can, found, has, made, some, they
14	a, at, down, her, his, if, may, one, or, runs, starts, who
16	away, eyes, first, hand, have, hear, see, that, through, us, we, world
18	an, needs, water
19	book, by, keep, new, read, things, when
20	children, he, not, people, she, states

Page	Content Words First Appearance
4	body, body systems, brain, nervous system, signals
5	communication, control, function, memory, organ
6	computer, messages, movement, problems
8	information, nerves, sensory organs
9	spinal cord
10	energy, halves, wrinkles
12	cells, glands, muscles, neurons
14	bundle, neck, paralysis, spine
16	ears, humans, nose, pain, scientists, senses, skin, stove, tongue, touch
18	B12, eggs, fish, meat, milk, vitamins
19	connections, instrument, puzzle, sport
20	Boston Children's Hospital, doctor, headaches, neurologist, United States, weakness

Published by AV2
14 Penn Plaza, 9th Floor New York, NY 10122
Website: www.av2books.com

Library of Congress Cataloging-in-Publication Data

Names: Noelle, Becky, author. | Das, Priyanka, author.
Title: Nervous system / Becky Noelle and Priyanka Das.
Description: New York, NY : AV2, [2021] | Series: My first look at body systems | Audience: Grades 2-3
Identifiers: LCCN 2020018432 (print) | LCCN 2020018433 (ebook) | ISBN 9781791118921 (library binding) | ISBN 9781791118938 (paperback) | ISBN 9781791118945 | ISBN 9781791118952
Subjects: LCSH: Nervous system--Juvenile literature.
Classification: LCC QM451 .N625 2021 (print) | LCC QM451 (ebook) | DDC 612.8--dc23
LC record available at https://lccn.loc.gov/2020018432
LC ebook record available at https://lccn.loc.gov/2020018433

Printed in Guangzhou, China
1 2 3 4 5 6 7 8 9 0 24 23 22 21 20

062020
100919

Project Coordinator: Priyanka Das Designer: Jean Faye Marie Rodriguez

The publisher acknowledges iStock and Shutterstock as the primary image suppliers for this title.